Tatjana Dmitrijewa
Elena Ivanova

Weather and Kelvin waves

Tatjana Dmitrijewa
Elena Ivanova

Weather and Kelvin waves

Weather and Kelvin waves

ScienciaScripts

Imprint

Any brand names and product names mentioned in this book are subject to trademark, brand or patent protection and are trademarks or registered trademarks of their respective holders. The use of brand names, product names, common names, trade names, product descriptions etc. even without a particular marking in this work is in no way to be construed to mean that such names may be regarded as unrestricted in respect of trademark and brand protection legislation and could thus be used by anyone.

Cover image: www.ingimage.com

This book is a translation from the original published under ISBN 978-620-0-44676-3.

Publisher:
Sciencia Scripts
is a trademark of
International Book Market Service Ltd., member of OmniScriptum Publishing Group
17 Meldrum Street, Beau Bassin 71504, Mauritius
Printed at: see last page
ISBN: 978-620-3-36594-8

CONTENTS

INTRODUCTION

The Earth's atmosphere is the gaseous envelope of our planet and extends up to thousands of kilometers above the planet's surface. It is characterized by high dynamics, physical heterogeneity, and susceptibility to biological factors. Over the billions of years in the history of the Earth's atmosphere, it was living organisms that primarily changed its composition.

The atmosphere is our protective dome against all kinds of threats from space. It burns up most of the meteorites that fall on the planet, and its ozone layer serves as a filter against the sun's ultraviolet radiation, the energy of which is deadly to living things. In addition, it is the atmosphere that maintains a comfortable temperature near the Earth's surface. Without the greenhouse effect, which is achieved by multiple reflections of sunlight from clouds, the Earth would be 20 to 30 degrees colder on average. The circulation of water in the atmosphere and the movement of air masses not only balance temperature and humidity, but also create an earthly diversity of landforms and minerals - such richness is found nowhere else in the solar system.

The mass of the atmosphere is 5.2×10^{18} kg. Although the gas shells extend many thousands of kilometers from the Earth, only those that rotate around the axis at a speed equal to the speed of rotation of the planet are considered the atmosphere. Thus, the height of the Earth's atmosphere is about 1000 kilometers, goes gently into space in the upper layer, the exosphere (from the other Greek "outer sphere").

This work also investigates the structure and properties of Kelvin-Helmholtz waves (VKG) in the atmospheric boundary layer (ABL) and a statistical generalization of the properties of VKG in different climate regions. Remote acoustic sensing of the ABL from the Earth's surface is used as the main method for identifying waves and quantifying the parameters of wave structures. Monostatic Doppler acoustic localizers (sodars), whose echo is caused by sound scattering in the presence of small turbulent air temperature irregularities, allow visualization of the mesoscale vertical ABL structure up to a height of several hundred meters and observation of waves in the range of refractive index variations and wind speed variations. Measurements of the frequency and amplitude properties of the echo signal from Sodar allow (with the inclusion of some accompanying measurements) to estimate the amplitudes and periods of waves, determine the structure of wave packets, their duration and height, estimate the speed and direction of their motion, observe the dynamics of generation and decay of waves. Internal gravitational waves (IGWs) with periods from ten seconds to ten minutes are characterized by a variety of shapes and properties associated with their origin and ABL layering. The ability of sodars to visually visualize the waveform allows gravitational shear waves generated in low-level jet streams to be distinguished from upwelling waves detected by near-surface waveguides. The main focus in the studies has been on Kelvin-Helmholtz type internal gravity shear waves, since they are the most common and are formed everywhere under inversion conditions, regardless of orographic obstacles, i.e., even over a smooth, uniform surface.

Relevance of the topic. The increase in the amount of anthropogenic pollutants in the atmosphere leads to an increase in demand for systems to monitor and predict the transport of pollution. The study of the mechanisms of metabolic processes occurring under different meteorological conditions is necessary to improve climate models and to model the dispersion of pollutants. In addition, metabolic processes in the APS play a critical role in the formation of weather and climate, affect the propagation of electromagnetic waves and sound waves, and influence wind energy efficiency and aviation safety. The development and operation of predictive numerical models require the parameterization of exchange processes in the

GSP. For the parameterization of an unstable and neutrally stratified ABL, the Monin-Obukhov similarity theory has been successfully used for decades, which allows to determine the necessary properties of turbulent exchange in the ABL using standard data on meteorological parameters. In a stably stratified ABL, despite numerous studies, the physical processes of mass and heat transfer remain poorly understood, poorly defined, and not amenable to universal parameterization. Surface layers and elevated inversions are known to be highly turbulent (as confirmed by remote sensing and optical aberrations in ground-based astronomical observations). However, the mechanism of turbulence generation during static stable stratification has not been fully explained.

Theoretical studies of internal gravity waves (divided into buoyancy waves (BWPs) and gravity shear waves (GDWs)) are extremely complex due to the nonlinearity of such wave motions. To date, not only are no analytical estimates of BWP and FGP parameters available, but there is also no generally accepted opinion on the degree of influence of wave activity of different types on the intensity of turbulence and exchange processes in a statically stable ABL. A variety of types of observed waves, the different conditions and means of observation do not even allow us to classify unambiguously different structures and to refer unambiguously to IGW of one or another type or to a particular type of frozen turbulence modulated as a result of wave or transition processes. Structures in the form of inclined stripes or ridges,

Kelvin-Helmholtz "under conditions where the vertical profile of the density of the medium is continuous).

To date, several hundred journal publications and separate chapters in special monographs have been devoted to internal gravity waves in the ABL. Nevertheless, the problems of VCG generation, determination of their properties and characteristics, determination of their relation to average meteorological parameters, and interaction with turbulence are far from being solved. Systematic experimental studies are needed to approximate these solutions.

The objective of this work is to investigate the parameters and internal structure of Kelvin-Helmholtz waves using data from long-term monitoring of the atmospheric boundary layer with acoustic localizers (sodars).

According to the set goal, the following tasks were solved:

1. Analyze the atmosphere of the planet Earth;

2. study vibrations in the atmosphere and Kelvin waves in the atmosphere ;

3. study of mathematical analysis and modern methods for the study of Kelvin waves and other atmospheric phenomena.

1. the atmosphere of the planet earth

Composition of the Earth's atmosphere. The history of the development

Although air seems to be homogeneous, it is a mixture of different gases. If we take only those that occupy at least one thousandth of the volume of the atmosphere, there are already 12. If you look at the general picture, the entire periodic table is in the air at the same time!

However, the Earth was not able to achieve such diversity immediately. It is only thanks to the unique coincidences of chemical elements and the presence of life that the Earth's atmosphere has become so complex. Our planet has preserved geological traces of these processes, so that we can look back billions of years:

1. the first gases that enveloped the young earth 4.3 billion years ago were hydrogen and helium - the basic components of the atmosphere of gas giants like Jupiter. These are the most elementary substances - the remnants of the nebula that gave birth to the Sun and the planets surrounding it were made of them, and they settled abundantly around the gravitational center planets. Their concentration was not very high and their low atomic mass allowed them to escape into space, which they still do. So far, their total specific gravity is 0.00052% of the total mass of the Earth's atmosphere (0.00002% hydrogen and 0.0005% helium), which is very low.

2. However, in the earth itself there were many substances that tried to escape from the hot gut. A large amount of gases were released from the volcanoes - mainly ammonia, methane and carbon dioxide, as well as sulfur. Ammonia and methane were subsequently decomposed into nitrogen, which now makes up the lion's share of the mass of the Earth's atmosphere - 78%.

However, the real revolution in the composition of the Earth's atmosphere occurred with the arrival of oxygen. It also appeared naturally - the glowing mantle of the young planet actively removed gases trapped under the Earth's crust. In addition, the water vapor emitted by volcanoes was split into hydrogen and oxygen under the influence of ultraviolet solar radiation.

However, this oxygen could not remain in the atmosphere for long. It reacted with carbon monoxide, free iron, sulfur and a variety of other elements on the surface of the planet - and high temperatures and solar radiation catalyzed chemical processes. This situation was changed only by the appearance of living organisms.

First, they released so much oxygen that they not only oxidized all the substances on the surface, but also accumulated - in a few billion years their amount increased from zero to 21% of the total mass of the atmosphere.

Second, living organisms actively used atmospheric carbon to build their own skeletons. As a result of their activities, the Earth's crust was filled with entire geological layers of organic materials and fossils, and carbon dioxide became much less

And finally, the excess oxygen formed the ozone layer, which began to protect living organisms from ultraviolet radiation. Life began to evolve more actively and take on new, more complex forms - highly organized creatures appeared among bacteria and algae. Today, ozone occupies only 0.00001% of the Earth's total mass.

You probably already know that the blue color of the sky on Earth is also produced by oxygen - of the entire rainbow spectrum of the sun, it best scatters the short wavelengths of light responsible for blue. The same effect works in space - at some distance the Earth seems to be shrouded in a blue haze, and from some distance it turns into a blue dot.

In addition, noble gases are present in significant quantities in the atmosphere. Among them, the most important is argon, whose share in the atmosphere is 0.9-1%. Its source is nuclear processes in the depths of the Earth, and it reaches the surface through microcracks in lithospheric plates and volcanic eruptions (in the same way helium occurs in the atmosphere).

Due to their physical properties, noble gases rise to the upper atmosphere and escape into space.

As we can see, the composition of the Earth's atmosphere has changed very much more than once and more - but this has taken millions of years. On the other hand, vital phenomena are very stable - the ozone layer will exist and function even if there is 100 times less oxygen on Earth. Against the background of the general history of the planet, human activities have not left serious traces. On a local level, however, civilization can cause problems - at least for itself. Air pollutants have already made life dangerous for residents of Beijing, China - and huge clouds of dirty fog over large cities are visible even from space.

Atmospheric structure

However, the exosphere is not the only special layer of our atmosphere. There are many of them, and each of them has its own unique properties. Let's take a look at some of the most important ones:

Troposphere

The lowest and densest layer of the atmosphere is called the troposphere. The reader of this article is now exactly in its "lower" part - unless he is one of the 500,000 people flying in an airplane right now. The upper limit of the troposphere depends on latitude (remember the centrifugal force of the Earth's rotation making the planet wider at the equator?) And ranges from7 kilometers at the poles to 20 kilometersat the equator. The size of the troposphere also depends on the season - the warmer the air, the higher the ceiling rises.

The name "troposphere" comes from the ancient Greek word "tropos", which translated means "to turn, change". This accurately reflects the characteristics of the atmospheric layer - it is the most dynamic and productive. In the troposphere, clouds gather and water circulates, cyclones and anticyclones are produced, and winds are generated - all these processes we call "weather" and "climate" take place. In addition, it is the most massive and dense layer - it accounts for 80% of the mass of the atmosphere and almost all of its water content. Most living organisms live here.

Everyone knows the higher you go, the colder it gets. It really is - every100 meters the air temperature drops by 0.5 to 0.7 degrees. Nevertheless, the principle only works in the troposphere - then the temperature rises with increasing altitude. The zone between the troposphere and stratosphere, where the temperature remains constant, is called the tropopause. And the wind flow also accelerates with altitude - by 2-3 km / s per kilometer upward. Therefore, para and hang gliders prefer plateaus and mountains for flights - they will always be able to "catch a wave" there.

The aforementioned air floor, where the atmosphere is in contact with the lithosphere, is called the surface boundary layer. Its role in the circulation of the atmosphere is incredibly large - the release of heat and radiation from the surface creates winds and pressure drops, and mountains and other irregularities in the relief direct and separate it. Every now and then there is an exchange of water - in 8-12 days all the water from the oceans and the surface returns, turning the troposphere into a kind of water filter.

An interesting fact is that an important process in the life of plants, transpiration, is connected with the exchange of water with the atmosphere. With its help, the flora of the planet actively influences the climate - for example, large green areas mitigate the weather and temperature drops. Plants in waterlogged places evaporate 99% of the water taken from the soil. For example, one hectare of wheat in summer releases 2-3 thousand tons of water into the atmosphere - this is much more than the lifeless soil could give.

The normal pressure at the Earth's surface is about 1000 millibars. The standard is the pressure of 1013 mbar, which represents an "atmosphere" - with this unit of measurement you have probably already bumped. With increasing altitude, the pressure drops rapidly: at the boundaries of the troposphere (at altitudes)12 kilometers) it is already 200 mbar and at an altitude 45 kilometersand falls to 1 mbar in total. Therefore it is not surprising that in the saturated troposphere 80% of all masses of the Earth's atmosphere are collected.

Stratosphere

The layer of the atmosphere ranging between 8 kilometer heights (at the pole) and 50 km(at the equator) is called stratosphere. The name comes from the other Greek word "stratos", which means "ground cover, layer". This is an extremely rarefied zone of the Earth's atmosphere where there is almost no water vapor. The air pressure in the lower part of the stratosphere is ten times lower than near the surface and 100 times lower in the upper part.

In our discussion of the troposphere we have already learned that the temperature in it decreases with altitude. In the stratosphere, everything happens in the opposite way - with the ascent, the temperature increases from -56 $^\circ$ C to 0-1 $^\circ$ C. In the stratopause, the boundary between strato- and mesosphere, the warming stops.

Ozone layer

And at the boundary between the stratosphere and the mesophera is the famous ozone layer. It protects the Earth's surface from the effects of ultraviolet rays and also serves as an upper limit for the spread of life on the planet - above which temperature, pressure and cosmic radiation will quickly put an end to even the most persistent bacteria.

Where does this shield come from? The answer is incredible - it was produced by living organisms, more precisely by oxygen emitted by various bacteria, algae and plants since time immemorial. Oxygen rises high into the atmosphere, comes into contact with ultraviolet radiation, and reacts photochemically. As a result, from the usual oxygen we breathe, O_2, ozone is obtained - O_3.

Paradoxically, ozone, which is produced by solar radiation, protects us from the same radiation! And ozone does not reflect, but absorbs ultraviolet light - thereby warming the atmosphere around it.

Mesosphere

We have already mentioned that above the stratosphere - more precisely above the stratopause, the boundary layer of stable temperature - is the mesosphere. This relatively small layer is located between 40-45 and 90 kilometers altitude and is the coldest place on our planet - in the mesopause, the upper layer of the mesosphere, the air cools down to -143 ° C.

The mesosphere is the least studied part of the Earth's atmosphere. Extremely low gas pressure, a thousand to ten thousand times lower than the surface pressure, limits the movement of balloons - their lift reaches zero and they simply hang in place. The same thing happens with jet airplanes - the aerodynamics of the airplane's wing and body become irrelevant. Therefore, either rockets or aircraft with rocket engines - rocket planes - can fly in the mesosphere.

And in the mesosphere, most meteors that fall to Earth burn out - this is where the Perseid meteor shower flares up , known as the "August Starfall" . The light effect occurs when a space body enters the Earth's atmosphere at an acute angle with a speed greater than 11 km / h - Due to the force of friction, the meteorite lights up.

After losing their mass in the mesosphere, the remains of the "aliens" settle on Earth in the form of cosmic dust - every day 100 to 10 thousand tons of meteorite matter fall on the planet. Since individual dust particles are very light, it takes up to a month for them to reach the Earth's surface! When they get trapped in the clouds, they make them heavier and sometimes even cause rain - just like volcanic ash or particles from nuclear explosions.

However, the impact of cosmic dust on precipitation is considered small - even 10,000 tons is not enough to seriously alter the natural circulation of the Earth's atmosphere.

Thermosphere

Above the mesosphere at an altitude of 100 kilometers above sea level runs the Karman line - the conditional boundary between Earth and space. Although there are gases that rotate with the Earth and technically enter the atmosphere, their amount above the Karman line is invisibly small. Therefore, any flight that goes above the altitude100 kilometersis already considered cosmic.

The lower limit of the most extensive layer of the atmosphere, the thermosphere, coincides with the Karman line. It rises to the height800 kilometers and is characterized by an extremely high temperature - at high altitude 400 kilometers it reaches its peak at 1800 ° C.!

Hot, isn't it? At a temperature of1538 ° C. Iron begins to melt - how then do spacecraft remain intact in the thermosphere? It's all about the extremely low gas concentration in the upper atmosphere - the pressure in the middle of the thermosphere is 1,000,000 lower than the air concentration at the Earth's surface! The energy of individual particles is high - but the distance between them is large and spacecraft are actually in a vacuum. However, this does not help them dissipate the heat given off by the mechanisms. To generate heat, all spacecraft are equipped with radiators that emit excess energy.

On a note. When it comes to high temperatures, it's always worth considering the density of the annealing matter - for example, scientists at the Andron Collider can actually heat matter to the temperature of the Sun. However, it is obvious that these will be separate molecules - one gram of the star's matter would be enough for a powerful explosion. Therefore, we should not believe the yellow press that promises us an immediate end of the world by the collider, just as we should not be afraid of the heat in the thermosphere.

Exosphere

The last layer of the Earth's atmosphere, whose lower boundary runs at a height 700 kilometers- This is the exosphere (from the other Greek root "exo" - outside, outside). It is incredibly scattered and consists mainly of atoms of the lightest element - hydrogen; also encounter individual oxygen and nitrogen atoms, which are strongly ionized by the all-pervading solar radiation.

The dimensions of the Earth's exosphere are incredibly large - it grows into the Earth's crown, the geocorona, which extends up to 100,000 kilometers from the planet. It is very rare - the particle concentration is millions of times less than the density of ordinary air. But when the Moon obscures the Earth for a distant spacecraft, the crown of our planet is visible when we see the crown of the Sun during its eclipse. However, this phenomenon has not yet been observed.

Survive the atmosphere

And in the exosphere, the Earth's atmosphere is weathered - due to the great distance from the planet's center of gravity, particles easily detach from the entire mass of gas and enter their own orbits. This phenomenon is called atmospheric dissipation. Our planet loses 3 kilograms of hydrogen and50 grams ofhelium from the atmosphere every second. Only these particles are light enough to leave the entire gas mass.

Simple calculations show that the Earth loses about 110,000 tons of atmospheric mass each year. Is it dangerous? No, the capacity of our planet to "produce" hydrogen and helium exceeds the rate of loss. Moreover, some of the lost matter eventually returns to the atmosphere. And important gases like oxygen or carbon dioxide are simply too heavy to leave the Earth en masse - so don't be afraid that the Earth's atmosphere will evaporate.

An interesting fact - "prophets" of the end of the world often say that the atmosphere quickly disappears under the pressure of the solar wind when the Earth's core stops rotating. However, our reader knows that gravitational forces keep the atmosphere close to the Earth, acting independently of the rotation of the core. A striking proof of this is Venus, which has a solid core and a weak magnetic field, but the atmosphere is 93 times denser and heavier than the Earth. However, this does not mean that the termination of the dynamics of the Earth's core is certain - then the planet's magnetic field disappears. Its role is less important in containing the atmosphere than in protecting it from charged particles of the solar wind, which easily turn our planet into a radioactive desert.

The clouds

Water on Earth exists not only in the vast ocean and numerous rivers. There are about 5.2×10^{15} kg of water in the atmosphere. It is present almost everywhere - the proportion of vapor in the air ranges from 0.1 to 2.5 vol .%, depending on temperature and location. However, most of the water is collected in clouds, where it is stored not only in the form of gas, but also in small droplets and ice crystals. The concentration of water in clouds reaches 10 g / m ^{3} - and since clouds reach a volume of several cubic kilometers, the mass of water in them is tens and hundreds of tons.

Clouds are the most visible formation on our Earth; they are visible even from the Moon, where the outlines of the continents blur before the naked eye. And this is not strange - after all, more than 50% of the Earth is constantly covered with clouds!

Clouds play an incredibly important role in the Earth's heat exchange. In winter, they trap the sun's rays and increase the temperature below them due to the greenhouse effect. In summer, they protect the sun's enormous energy. Clouds also balance the temperature differences between day and night. Incidentally, it is because of their absence that deserts cool so much at night - all the heat accumulated by sand and stones flies freely upward as clouds hold it in other regions.

The vast majority of clouds form near the Earth's surface in the troposphere, but take on a variety of shapes and characteristics as they develop. Their separation is very useful - the appearance of different cloud types can not only help predict the weather, but also determine the presence of contaminants in the air! Let's take a closer look at the main cloud types.

Lower clouds

The clouds that are lowest above the ground are called lower level clouds. They are characterized by high uniformity and low mass - when they fall to the ground, meteorological scientists do not separate them from normal fog. Nevertheless, there is a difference between them - some simply cover the sky, while others can erupt into large rainfalls and snowfalls.

Stratus clouds belong to the clouds that can cause heavy precipitation. They are the largest among the lower level clouds: their thickness reaches several kilometers and their linear dimensions exceed thousands of kilometers. They are a homogeneous gray mass - look at the sky during prolonged rain and you will surely see stratus clouds.

Another type of low-level clouds are stratocumulus clouds, which rise 600 to 1500 meters above the ground. They are groups of hundreds of gray-white clouds separated by small gaps. We usually see such clouds on cloudy days. It rarely rains or snows from them.

The last type of lower clouds are ordinary stratus clouds; they cover the sky on cloudy days when a drizzle comes from the sky. They are very thin and low - the height of stratus clouds reaches maximum 400-500 meters. Their structure is very similar to that of fog - at night they sink to the ground and often create a dense morning haze.

Vertical development clouds

The lower level clouds have older brothers - the clouds of vertical development. Although their lower limit is at a low altitude of 800 to 2000 kilometers, clouds of vertical development race upward in earnest - their thickness can reach 12 to 14 kilometers, which pushes their upper limit toward the limits of the troposphere. Such clouds are also called convective: Due to their size, the water in them assumes a different temperature, which creates convection - the process by which hot masses move upward and cold ones move downward. Therefore, in the clouds of vertical development simultaneously exist water vapor, small droplets, snowflakes and even whole ice crystals.

The main type of vertical clouds are cumulus clouds - huge white clouds resembling torn pieces of absorbent cotton or icebergs. For their existence high air temperature is necessary - therefore in central Russia they appear only in summer and melt at night. Their thickness reaches several kilometers.

However, when cumulus clouds can come together, they form a much larger shape - cumulonimbus clouds. From them in the summer come violent showers, hail and thunderstorms. They exist only for a few hours, but at the same time become adult15 kilometers- Their upper part reaches a temperature of -10 ° C and consists of ice crystals. On the tops of the largest cumulonimbus clouds form "anvils" - flat areas resembling a mushroom or an inverted iron. This happens in the areas where the cloud reaches the boundary of the stratosphere - physics does not allow it to spread further, so the cumulonimbus cloud spreads along the altitude boundary.

An interesting fact is that powerful cumulonimbus clouds form in places like volcanic eruptions, meteorite impacts and nuclear explosions. These clouds are the largest - their boundaries even reach the stratosphere and rise to a height16 kilometers.... Being saturated with evaporated water and microparticles, they erupt heavy thundershowers - in most cases this is enough to extinguish the fires associated with the disaster.

Medium clouds

In the middle part of the troposphere (at a height of 2 to 7 kilometers in mid-latitudes) there are clouds of medium height. They are characterized by large areas - they are less affected by updrafts from the Earth's surface and uneven terrain - and a small thickness of several hundred meters. These are the clouds that "wrap" around the sharp peaks of mountains and hang near them.

The middle clouds themselves are divided into two main types - altostratus and altocumulus.

Altostratus clouds are one of the components of complex atmospheric masses. They are a uniform gray-blue veil through which the sun and moon are visible - although highly stratified clouds are thousands of kilometers long, they are only a few kilometers thick. The dense gray shroud visible from the window of an airplane flying at high altitude is precisely those highly stratified clouds. Often they have long rains or snow.

Altocumulus clouds, resembling small pieces of torn absorbent cotton or thin parallel stripes, are found in the warm season - they form when warm air masses reach a height of 2-6 kilometers. Altocumulus clouds serve as a sure indicator of the upcoming weather change and the approach of rain - they can be produced not only by natural convection of the atmosphere, but also by the onset of cold air masses. It rarely rains from them - clouds, however, can collide and form a large rain cloud.

Speaking of clouds near the mountains - in the photos (and maybe even live) you have probably seen more than once round clouds resembling cotton pads hanging in layers above a mountain top. The fact is that mid-layer clouds are often lenticular or lens-shaped - divided into several parallel layers. They are formed by air waves created when wind flows around steep peaks. Lenticular clouds are also special in that they hang in place even in the strongest winds. This is made possible by nature - since such clouds are formed at the contact points of several air currents, they are in a relatively stable position.

High clouds

The last level of ordinary clouds rising to the lower reaches of the stratosphere is called the upper level. The height of such clouds reaches 6-13 kilometers - it is very cold there, and therefore the clouds on the upper level consist of small pieces of ice. Because of their fibrous, stretched shape, resembling feathers, high clouds are also called cirrus - although the quirks of the atmosphere often give them the shape of claws, flakes and even fish skeletons. The precipitation that forms from them never reaches the ground - but the presence of cirrus clouds serves as an ancient method of predicting the weather.

Cirrus clouds are the longest of the upper level clouds - the length of a single filament can reach tens of kilometers. Since the ice crystals in the clouds are large enough to feel the Earth's gravity, cirrus clouds "fall" in whole cascades - the distance between the upper and lower points of a single cloud can reach 34 kilometers! In fact, cirrus clouds are huge "icefalls". It is the differences in the shape of the water crystals that create their fibrous, flowing form.

This class also includes practically invisible clouds - cirrostratus clouds. They form when large amounts of near-surface air rise upward - at high altitudes, their moisture content is sufficient to form a cloud. When the sun or moon shines through them, a halo appears - a luminous rainbow disk of scattered rays.

Noctilucent clouds

Noctilucent clouds, the highest clouds on Earth, should be divided into a separate class. They climb high80 kilometersthat's even higher than the stratosphere! They also have an unusual composition - unlike other clouds, they consist of meteorite dust and methane, not water. These clouds are visible only after sunset or before sunrise - the sun's rays penetrating from the horizon illuminate noctilucent clouds that remain invisible at altitude during the day.

Noctilucent clouds are an incredibly beautiful sight - but special conditions are required to see them in the northern hemisphere. And their mystery was not so easy to solve - the fainting scientists refused to believe in them and explained the silver clouds as an optical illusion.

2. oscillations in the atmosphere and Kelvin waves in the atmosphere.

There are many different types of atmospheric internal gravity waves. In the atmospheric boundary layer, the most common waves observed are those caused by the instability of the airflow during vertical shear of the wind speed. Such waves are recorded everywhere with stable temperature stratification, even in the absence of orographic obstacles or irregularities. It is these gravitational shear waves that are the focus of this paper.

Physical description of the VKG

The existence of mesoscale wavy structures in the lower troposphere (with wavelengths ranging from hundreds of meters to several kilometers) was already known in the 19th century through observations of "swirling" clouds. Examples of photographs of such clouds with characteristic swirling in the upper part are shown in the figure.

Number: 1.1. examples of photographs of the Kelvin-Helmholtz shafts on top of the clouds. a) Taiwan, photo by Jochun Ho; b) Cape Fiolent, photo by Taisiya Lazareva.

Apparently, such structures form due to the instability of the air flow with a jump in density and wind speed at the upper cloud boundary and are a typical example of Kelvin-Helmholtz waves (VCG), named after scientists who theoretically studied the formation process about 150 years ago waves and lambs on the water surface under the influence of wind. Subsequent theoretical work [36; 37; fifty; 76; 77; 99] discovered a number of properties of atmospheric VCG:

- found a criterion for the necessary condition for the existence of VCG - the critical value of the Richardson number;

- It is shown that the wavelength and the type of motion with the maximum linear growth rate depend on the profiles of the average stability velocity.

- It is shown that the phase velocity of the VCG motion is close to the average velocity of the medium, from which it follows that the observed wave structures of the cat's eye or braid type are vortex formations (or vortex chains). in the region of shear instability rather than propagating waves.

Moreover, it was hypothesized that the collapse of the VCG is the main cause of turbulence and vertical mixing in static stable media [45]. This hypothesis implies the important role of the VCG in the processes of turbulent exchange in the atmosphere and the need to consider the VCG in the parameterization of such processes. As a result, a variety of theoretical models for shear flows have been developed, as well as laboratory demonstrations of the formation and evolution of Kelvin-Helmholtz waves.

It should be noted that neither in Russian nor in foreign literature there is an established terminology for designating such wave motions in a cloudless atmosphere. In many foreign publications the considered structures are called Kelvin-Helmholtz waves (VKG). In the rest of the text, this term and abbreviation will also be used when describing internal gravitational waves whose source is the shear instability of the flow. It is important to note that the classical Kelvin-Helmholtz waves are only a special case of gravitational waves in shear flows, realized with a discontinuity in the density and velocity of the medium. In the troposphere, such abrupt discontinuities do not usually exist,

It is now generally accepted that VCGs are formed as a result of a loss of flow stability under the influence of vertical shear of the horizontal flow velocity in a statically stable medium. In this case, energy is transferred from the mean flow to the wave and later, when the wave breaks, to small turbulent fluctuations. The question of the reverse effect of small fluctuations in the shear flows on the loss of stability remains open. Quasi-periodic structures may form with the development of shear instability. In remote sensing of the atmosphere with sodars, radars, or lidars, these structures leave a characteristic pattern on the time base of the echo. Schematic examples of such echograms are shown in the figure.

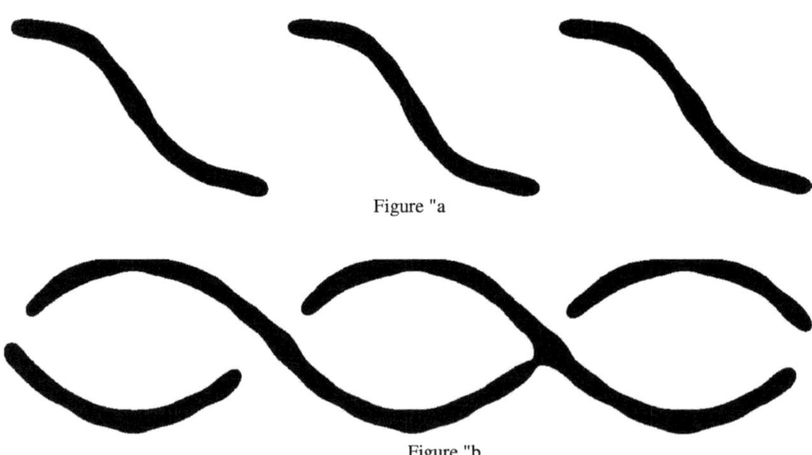

Figure "a

Figure "b

Number: 2. Schematic representations of Kelvin-Helmholtz waves observed on time bases of the echo signal during remote sensing of the atmosphere. a) - "strings" or "inclined stripes" (pigtails) - the most common type of VKG recorded with sodars and radars; b) closed structures, the so-called. "cat's eyes" (cat's eyes).

In the 1970s and 1980s, about a dozen different episodes of VCG observation in the ABL were described, including along with measurements of turbulence features [25; 27; 41; 57; 85]. The interaction with turbulence in these papers was discussed only at the qualitative level. Laboratory studies of waves in shear flows in two-layer and three-layer fluids began in the same years [100; 101] and continue to this day [71; 86]. The research results have shown good agreement both with the wave-like structures observed in clouds and with the results of the numerical simulation of the VKG [2].

The last decade has seen an increase in the number of observations of various types of IGW in the ABL using lidars [23; 24; 88]. Despite the intensive development of ground-based remote sensing equipment, the vast majority of papers published on this topic in recent decades have been devoted to either the mathematical aspects of internal gravity wave theory or the numerical analysis of hypothetical models [see, e.g., 15; 17; 19; 31; 45; 73; 87; 105]. Experimental studies of VKG in the lower troposphere have been an order of magnitude smaller than theoretical ones.

Continuous long-term sodar measurements of the parameters of the atmospheric boundary layer, carried out at the A.M. Obukhov RAS enabled the creation of a comprehensive catalog of cases of the occurrence of internal gravity shear waves in the atmospheric boundary layer. In the following statistical data, historical data of measurements performed over 3 years2009 year... by2011 year... are used. [12].

To date, a number of algorithms have been developed and successfully implemented to automatically detect wave structures on soda echograms. However, for certain sodars and measurement conditions (especially in a highly noisy conurbation), much additional work is required to adapt and test the automatic detection. The method tested in [7], based on the Haar cascade classifier, showed good results in identifying waves in the test sample. However, it requires "training" on fragments of images of echograms with wave registration, so the shape of the structures identified by the program is determined by sample selection.

Despite the possibility of refinement and development, this method is difficult to apply in conditions of strong variability in the shapes of the observed wave structures, and most importantly, it is impossible to use it to distinguish between BVP and different types of VCG. Taking this into account, traditional visual identification of VCG episodes was performed on echograms. In the analysis of soda data, VCG episodes in echograms were distinguished independently by two operators using the following criteria:

The minimum period of wave-like structures exceeds the period of transmission of the probe pulses (i.e., The temporal resolution of the sodar) by at least 12 times (> 3 minutes).

The thickness of the layer of wave activity (i.e., twice the amplitude of the wave) exceeds the height resolution of Sodar by at least three times (> 60 m).

The modulation depth of the background level of the echo signal by the wave structure exceeds 5 dB.

A packet of wavy structures contains at least 3 wavy periods.

Identification of structures with shorter periods, duration, amplitude and modulation depth is unreliable due to the limited temporal and spatial resolution of the sodar. LATAN-3 allows the adjustment of radiation parameters (carrier frequency, duration and repetition period of sampling pulses, sensitivity, etc.) over a wide range. However, these parameters have been set to the above values based on the multifunctionality of the APS monitoring mode on the CSS. Criteria 1) - 5) limit the number of VCGs, and smaller and shorter waves were not included in the statistics.

The structure of the VCG is clearly visible there The modulation of the echo signal intensity by "stripes" reaches 15-20 dB. In addition to such structures, many less clear VCGs were found, where the modulation depth was only 5-10 dB. In the statistics, there is no division of cases by modulation depth, since such a division did not lead to noticeable changes in the statistical distribution of wave parameters. For 3 years of measurements around the clock, 234 cases (points connected by lines) and wind directions in points (points without lines) were selected. Wind speed values with low statistical confidence are indicated by lines without points.

VKG with a duration of wave trains from 15 minutes to several hours, according to the specified criteria. The total duration of the episodes was about 400 hours.

Most VCGs were observed in January and September and the smallest in February, October, and December. However, the interannual variability of the amount exceeds the interseasonal variability, so the presented distribution does not allow to draw a definitive conclusion about the annual course of the frequency of occurrence of VCGs at the CSS. Apparently, their occurrence is not directly related to the average air temperature and the direction and speed of the wind, which show seasonal variability.

The average period of the VKG bands for each case was determined visually from the echogram. The distribution shows a rapid decrease in the number of repetitions of cases with periods longer than 3 minutes. At the same time, several VCG episodes with anomalously long periods were detected, corresponding to wavelengths of 5-.8 kilometers.... It is obvious that the asymmetry of this empirical distribution is caused by the choice of criteria for VCG selection.

Using Taylor's "frozen turbulence" hypothesis, an estimate for the observed wavelength can be obtained as the product of the period determined from the echogram and the average wind speed in the layer with VCG. This parameter allows a qualitative comparison with the results of the VCG model concepts. According to Howard's semicircle theorem [56], a necessary condition for the instability of the layer is the criterion for the

Richardson number: Ri <1/4 in a region within the layer. Then gravitational shear waves can occur in the medium with a positive growth rate as a function of the wave number and the Richardson number. The range of values that the ratio of the length of the internal gravitational shear wave L to the thickness of the layer of wave activity h can assume depends on the type of model for height changes of the density and velocity of the medium. and also on Richardson's number. Thus it is possible to investigate (similarly

[26]) To what extent are the measurements consistent with theoretical models and laboratory and numerical experiments investigating Kelvin-Helmholtz instability? Early theoretical studies showed that the wavelength and type of motion with the maximum growth rate depend on the velocity and stability profiles and the presence of a hard lower bound [36; 76]. Typical wavelengths range from L = 2nh to 7.5h [77]. The depth to which the layer is mixed also depends on the properties and initial degree of flow instability. Laboratory and two-dimensional numerical studies have shown that the maximum amplitude depends simultaneously on the Richardson number and the Reynolds number, with large rolls (and a deeper mixed layer) corresponding to a lower Ri and a higher Re [100]. These results have been confirmed and extended during recent three-dimensional modeling studies of CG instability and transition to turbulence [105]. In particular, studies have shown the inverse dependence of the amplitude of the rolls and the thickness of the mixed layer on Ri at a fixed L.

The accuracy of the visual determination of the period and amplitude of the wave is limited by the resolution of the sodar. The average flow rate was calculated as the arithmetic mean of the maximum and minimum values in the layer. The average L / h ratio calculated for 234 cases was 9, 2, which is comparable to the L / h = 7.5 value corresponding to the fastest growing mode in the model with linear velocity and density profiles [77].

To clarify the relationship between the magnitude of wind shear and the occurrence of VCH, the displacements in the lower part of all NSTs observed at the ZSS in 2009 and the flows where VCHs were observed were calculated separately. In the literature, there is no generally accepted method for assessing the magnitude of wind speed shear in the layer because the shapes of the vertical velocity profiles vary widely. In this work, wind shear in the layer was calculated as the difference between the maximum and minimum values of wind speed in the layer, with respect to the layer thickness. Shows the distribution of the total duration of VCH at different wind shears for all cases for 2008-2011, as well as the distribution for the one-hour averaged wind shears in the nighttime NST for 2009, calculated from the echograms. The distribution shape for cases with VCG shows no discernible differences:100 mand with slight asymmetry to large shifts. Thus, the observations showed no direct (i.e., Excluding temperature gradient) influence of wind speed shear on the occurrence of VCG.

Overall, the results presented provide fairly comprehensive statistics of the distributions of VCG parameters. Basically, waves were observed in the form of "oblique stripes" at an altitude of 100-.300 m, with wind shear 3-5 m / s at 100 m.... Typical parameters of the observed waves: Wavelength 400-4000 m, double amplitude (thickness of the layer of wave activity) 60-300 mmeans the average ratio between the wavelength and the layer thickness $L / h = 9.2$.Estimates for the layer thickness and period are consistent with the statistics for 72 episodes of IGW recording in echograms obtained by the method of automated identification of waves and their parameters.

The lifetime of VKG packets was usually between one and several hours. In summer, wave activity was sometimes observed continuously throughout the night. In winter, when long-term inversions, sometimes lasting several days, form over the snow surface, the lifetime of VCG trains reached 10 to 15 hours. During synchronous observations of the VCG, various wind speed directions were recorded, including perpendicular to the Moscow-ZNS direction. Such observational results indicate that regional synoptic conditions play a key role in the occurrence of internal gravity shear waves and outweigh local features of the observation area.

Comparisons of typical regional episodes of VKG (with simultaneous registration at the Moscow State University and ZSS points) from the archive of soda measurements in the Moscow region for 2012-2014 were performed. with synoptic maps based on reanalysis data. Situations with high pressure (anticyclonic conditions), low pressure (cyclonic conditions) and atmospheric fronts in the measurement area were considered separately.

In anticyclones, there is a regular daily evolution of the mesoscale turbulent structure of the ABL. Nighttime radiative cooling produces surface inversions where small jet streams with large wind shears are observed; after sunrise, the thickness of the inversion layer begins to increase, followed by its rise and destruction and transition to convection. Despite the large gradients of wind speed at night, the VCG is rarely recorded in anticyclones: The nighttime surface inversion layer is usually too thin to resolve its structure with standard sodars. Examples of recording such small VCGs with a high-resolution minisodar were given in [21]. Also, the absence of waves can be explained by the fact that for the strongly stable stratification of the nighttime ABL, which is common for anticyclones, the Richardson number exceeds the critical value Ricr = 0.25, i.e., the necessary condition for the occurrence of the Kelvin-Helmoltz instability in the flow is violated: Ri <Ricr [48].

At the same time, VCGs are almost always observed during the morning rise of the inversion for 2-3 hours before the destruction of the inversion layer. The time of wave registration may be different for different observation points in the same region due to various factors affecting the rise time of the inversion layer: the nature of the underlying surface, the influence of the urban heat island. Under the conditions of anticyclones, there are no extensive VCG trains on a regional scale with a duration of more than three hours.

Under low pressure conditions, VCGs are recorded relatively infrequently. Cyclones are usually characterized by cloudy skies with no low-level jet streams developing, where shear waves are usually formed. In addition, precipitation is observed most of the time, and soda data for these episodes are highly noisy. Rare intense trains of gravity shear waves, as well as traveling waves or breaking Kelvin-Helmholtz waves, may be observed under cyclonic conditions, with fronts accompanied by large gradients of meteorological values.

Extended and long-term VCG trains are usually observed during large surface pressure gradients at the boundaries of baric formations or during the passage of atmospheric fronts. At the boundaries of anticyclones and under transitional conditions, extended VCG trains lasting up to 10 hours or more are regularly recorded during clear weather (observed throughout the existence of the nighttime surface inversion). Two different factors are possible to explain the relationship of wave activity in the ABL to such synoptic conditions:

- Due to the moderate turbidity and weak static stability of the ABL, Richardson numbers often assume values lower than the critical value, and Kelvin-Helmholtz instability occurs, leading to the development of VCG.

- Due to the baroclinicity of the troposphere, traveling gravity waves are generated therein, which are partially trapped by the surface waveguide present in the jet stream. Such trapped waves appear as standing fluctuations of the air density in the waveguide, similar to Kelvin-Helmholtz shear waves.

The influence of atmospheric fronts on the generation of tropospheric inertial gravity waves was recently studied in [90]. Obviously, the main conclusions of this work can also be attributed to VKG in the atmospheric boundary layer.

These are colmgor spectra of temperature and wind speed variations for locally homogeneous and isotropic turbulence. Structural properties of temperature and wind speed.

$$C_T^2 = \langle (T_1 - T_2)^2 \rangle r_{1,2}^{-2/3}, \, C_v^2 = \langle (v_1 - v_2)^2 \rangle r_{1,2}^{-2/3},$$

("Law 2/3" of Kolmogorov-Obukhov), T is the average air temperature, c is the average speed of sound, $r_{1.2}$ is the distance between the temperature measurement points T_1 and T_2, k = 2p / L, K = 2k sin (0/2) = 2n / lt is the wavenumber of turbulence on a linear scale lt [59].

From the formula it can be seen that at a scattering angle of 90 degrees no scattering occurs and at 180 degrees only temperature variations contribute.

For scattering angles where the factor $\cos^2$ (0/2) is close to one, the contribution of temperature variations to the intensity of the scattered signal is usually 1-2 orders of magnitude smaller than the contribution of wind variations. For backscatter (scattering at an angle of 180 degrees) used in monostatic sodar, the effective intercept is given by the first term in the expression:

$$\sigma_{180} = 2\pi k^4 \Phi_T(K)(2T)^{-2},$$

K = 2k is the K value for the backscatter.

$$\sigma_{180} = 3,9 \cdot 10^{-3} k^{1/3} C_T^2 T^{-2} = 7,5 \cdot 10^{-3} \lambda^{-1/3} C_T^2 T^{-2},$$

$$C_T^2 = 140 \sigma_{180} \lambda^{1/3} T^2.$$

The formulas are correct provided that the characteristic scale of scattering inhomogeneities, $lt = A / 2$, lies in the "inertial turbulence interval" in which the hypothesis of homogeneous and isotropic turbulence is applicable. As can be seen from the Kolmogorov spectral plot shown in the figure, this condition is satisfied for sodars with reference frequencies from 1.5 to 6.0 kHz, which corresponds to $L / 2 = 2/10$ cm.

The backscatter value is expressed as the ratio of the emitted and received sound power Pt and Pr for monostatic sodar by the following formula:

$$\sigma_{180} = P_r R^2 \exp(2\alpha R)[P_t(c\tau/2)S_{ant}]^{-1},$$

Where: R is the distance between the sound antenna and the scattering volume, a is the absorption coefficient of sound in air, st / 2 is the thickness of the scattering volume (c is the speed of sound, t is the duration of the sound pulse), Sant is the effective area of the antenna. The attenuation coefficient a is determined by the mechanism of molecular absorption of sound in air and depends on the air temperature, humidity, pressure and frequency of the sound signal. For the echo signal of a monostatic sodar with a reference frequency of 2 kHz, the following estimates of the quantities contained in the expression can be given (at air temperature T = 273 K, relative humidity h = 70%, and pressure p = 1013 mbar, and at typical values $C^2{}_T = 10^{-2}/10^{-4}$ K^2 m$^{-2/3}$ for the sound layer at an altitude of 100 m [61]:

a = 10-8 -т 10-10 м-1;

Pt = 10 Вт;

a = 1.4 дБ/100 м;

т = 0.1 с;

Pr = 10-11 - 10-13 Вт.

The scattering inhomogeneities are carried by the wind (according to the frozen turbulence hypothesis), therefore it turns out that the frequency of the reflected signal f is shifted relative to the frequency of the sound pulse fo due to the Doppler effect. Thus, the wind speed component parallel to the sound direction is expressed by the following formula:

$$V_r = \frac{c}{2f_0}(f_0 - f).$$

In the cloudless atmospheric boundary layer (ABL), wave structures were not observed until the advent of ground-based remote sensing instruments (sodars, radars, and lidars), although microbarographic studies of IGW in the lower and upper troposphere have been successfully conducted in many countries. It is impossible to reconstruct the structure and height of the wave packet from periodic pressure fluctuations of a wave-like nature, so VKGs could not be isolated from the general set of IGW observations of various types. Later, when microbarographic recordings were performed simultaneously with acoustic sound, pressure fluctuations were recorded during the occurrence of wave-like interlayer structures [55].

Volumetric scattering of sound and radio waves recorded by sodars and radars occurs on a small scale (with dimensions corresponding to half the wavelength of the probe radiation and several centimeters) turbulent inhomogeneities of the refractive index [13; eighteen; 58]. The agreement of IGW indication with echograms and real wave structures has been confirmed by laboratory experiments [71; 86], as well as registration of wavy structures of the same shape in the wind speed field and wind speed shear field [28; 82], and at present there is no doubt.

Most observations of atmospheric VCG have been made at altitudes of 2-20 km and 50-100 km in the course of regular radar surveys of the troposphere, lower stratosphere, and ionosphere. This is due to the proliferation of networks for continuous monitoring of the atmosphere by pulsed microwave and radio frequency radars, whose vertical resolution is (100-).200 m) is not sufficient to study a stably layered ABL. Detailed statistics on the properties and parameters of the VCG at these altitudes can be found in reviews [47; 79]. There are few VCG observations in the atmospheric boundary layer, apparently explained by the lack of network monitoring of the ABL and the episodic nature of remote sensing of this layer. From 1968 to the present, only about a dozen and a half detections of VCG packets in the APS have been published with frequency-modulated radars, sodars, and lidars that have sufficiently high vertical resolution (2-)20 m) for the study of wavy formations, and only two studies performed comparisons of the results of VCG observations with theoretical models, which were qualitative rather than quantitative [26; 34; 39; 40; 82].

For each case, the values of the height of the center of the wave layer, the average vertical wind shear in the layer, the thickness of the wave layer, and the wavelength (or spatial period of the structures calculated on the basis of the Taylor hypothesis) are given. The ratio of the wavelength to the layer thickness is also presented, which allows comparison with the results of theoretical studies and data from numerical and laboratory experiments .

Shows the samples of the first registrations of Kelvin-Helmholtz waves in a cloudless APS on radar and soda echograms at coordinate height - current time [30; 51; 52; 60]. The degree of blackness in the diagrams corresponds to the degree of intensity of the signal received by the tracking device caused by the backscattering of radio waves and sound at the inhomogeneities of the refractive index in air. Since the intensity of the received signal is directly proportional to the structural properties of turbulent temperature variations of (for sodar) or humidity C_2 (for radar)

[18], the darker the image of the structure in the echogram, the more turbulent it is. Trains of wavy structures are clearly visible on the echograms.

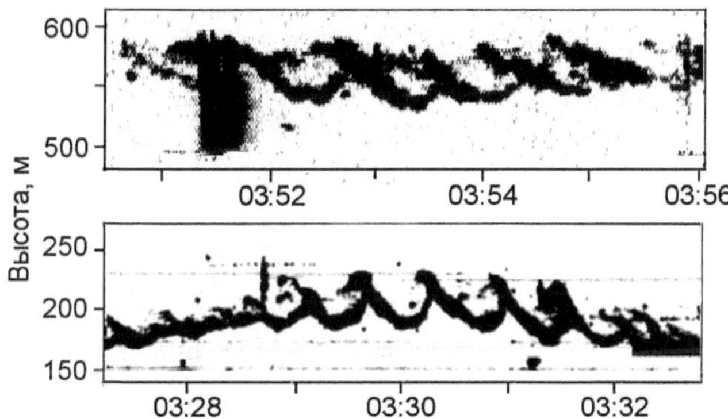

In Fig. 1.3 (a) shows VCGs in the form of "cat's eyes" and "rolls" (see Fig. 1.2 (a)) recorded by FM radar in a thin layer of enhanced inversion.

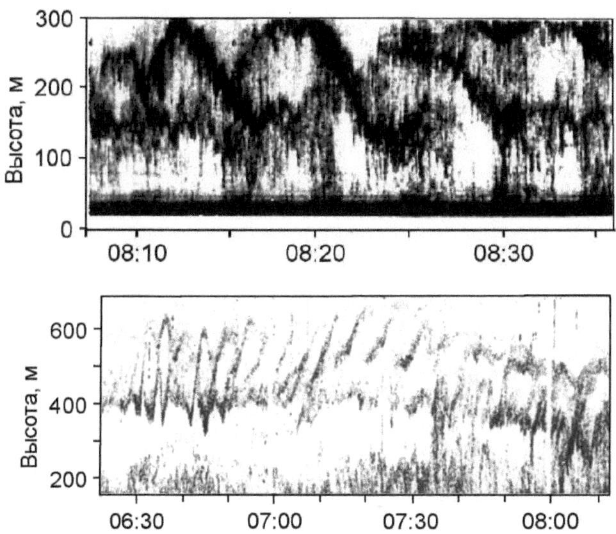

In Fig. 1.3 (b) and 1.3 (c), soda echograms with VCG show "cat's eyes" and "oblique stripes" in the layers of raised inversions.

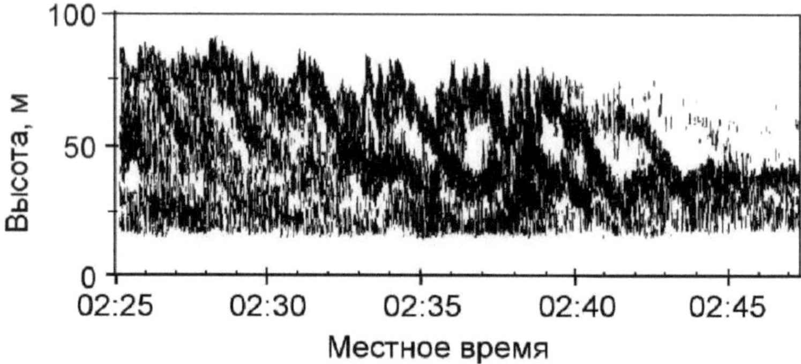

In Fig. 1.3 (d) shows the "cat's eyes" in the surface inversion layer obtained using a high spatial sodar (3.5 m) and time resolution (1 s).

The heights of the VCGs, their amplitudes, and periods differ markedly from each other, but the shapes of these wave formations occupy the entire thickness of the layer elevated or surface temperature inversion, easily discernible. They do not at all resemble the nearly sinusoidal shape of traveling internal gravity waves (buoyancy waves), which are also occasionally recorded on echograms in the form of variations in the height of the boundaries of the inversion layers.

The image of the contour of moving buoyancy waves on the echograms of sodars, radars and lidars looks like a unique function of current time or (for scanning devices) distance, and the image of the contour of Kelvin-Helmholtz waves looks like a two-digit function. VCG images similar to those were also obtained by the author during sodar measurements at Zvenigorod Scientific Station (ZSS) and under expedition conditions in 2007-2018.

Shows examples of VGC visualizations in horizontal expansion height coordinates obtained using a scanning Doppler lidar and a high-resolution scanning pulse radar. Spatial structure in the form of rolls is clearly visible in the radial wind speed field in the surface inversion layer (Fig. 1.4 (a)); "cat's eyes" were observed in the vertical shear field of horizontal wind speed in the raised inversion layer (Fig. 1.4 (b)). Such spatial images allow to estimate the horizontal length of VCR packets and wavelength without resorting to the hypothesis of a "frozen" Taylor turbulence.

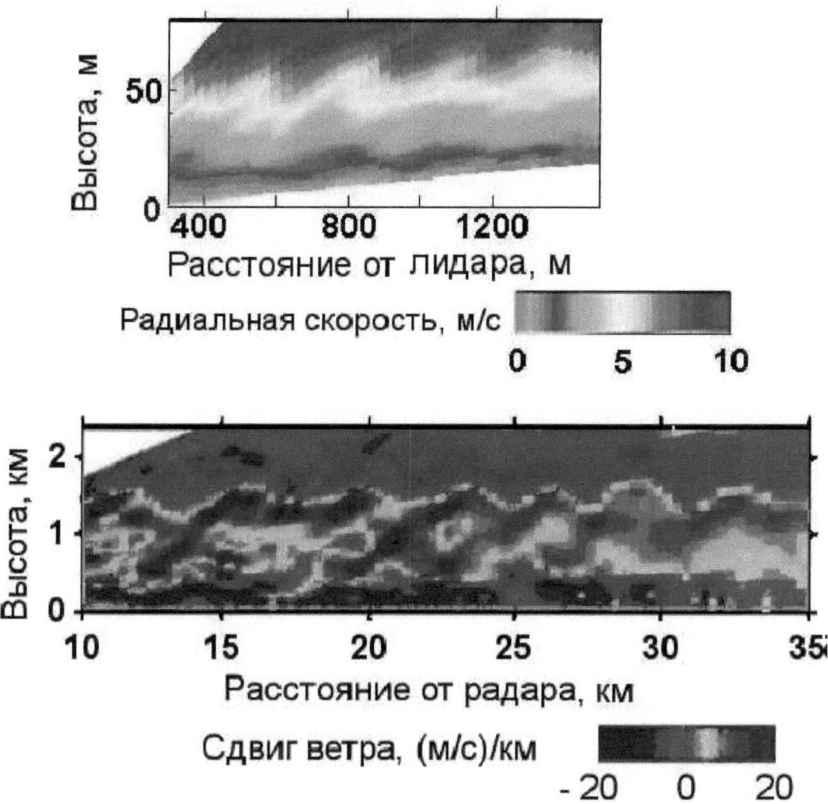

Number: 1.4. Examples of VCG visualization on echograms in horizontal longitude-height coordinates obtained with (a) a scanning Doppler lidar [82] and (b) a high-resolution scanning pulse radar [28].

Sodar is the simplest and least expensive instrument of all ground-based remote sensing instruments suitable for visualization and study of VCG in the atmospheric boundary layer [29]. Modern high-frequency mini-sodars with high spatial resolution (ca.2 m) [89] allow detailed study of wave processes in the lower part of the ABL.

Of particular importance are studies of VCG in the atmospheric boundary layer, since waves can have a significant effect on heat and mass transfer. In addition, studies of waves in the ABL are needed to solve many applied problems associated with short-range transport of pollution and with wind energy [see, for example, 63].

In the last decade, much attention has been paid to the interaction of waves in ABL with turbulence [see, for example, 7; 14; 33; 70; 83; 91; 97; 98]. This is mainly due to the development of large-scale regional modeling of ABL, which requires appropriate parameterization of metabolic processes with stable stratification [92]. All these works have one thing in common: one or two episodes of observation of VKG packages are studied in detail.

Different VCG parameters are observed in different episodes, and the comparison is made with different ABL properties (including undulations of temperature and wind speed variations). Therefore, it is difficult to draw a general picture of the interaction from these studies. Several hypotheses have been proposed to explain the formation of periodic rolls in the turbulent structure of shear flows.

One of the explanations is the generation of turbulence during the destruction of Kelvin-Helmholtz waves resulting from shear instability. Another mechanism may be the capture of traveling waves from orographic obstacles by the surface waveguide. Periodic structures can also be explained by the collapse of traveling waves and the subsequent deformation and transmission of a turbulent wake by a shear flow.

The physical mechanisms have not yet been systematized and no quantitative relationships have been established between IGW parameters and external parameters. Moreover, since many of these papers do not use visualization of wave motion with sodars, lidars, or radars, it is impossible to understand what form of VCG (ripples, "cat's eyes," "fish bones") was observed and whether it was VCG at all or trapped internal gravity waves modulating the height of the inversion layer boundaries. Reviews on this subject constantly note the difficulties of theoretical studies of stable ABL, including the particular difficulty of studying the nonlinear problem of the interaction of wave activity and turbulence [1; 43; 54; 74].

Another widely used direction of experimental studies on the influence of VCG on the intensity of turbulence is the development of methods for evaluating the relationship between waves and turbulence, which are practically important for ensuring aviation safety, but are not very promising for clarifying the physical interaction processes. These methods are based on works [65; 66] using an indirect relationship between synoptic situations where the probability of occurrence of VCG is high and aircraft observations of "clear air turbulence". In works [38; 62] analyzed the synoptic situation in many commercial airline accidents and used passenger aircraft pilots' reports of areas of high turbulence. According to estimates in Lit. [104] by numerical simulation

The main part of recent publications on the problem of VKG in the atmosphere is aimed at mathematical modeling of atmospheric waves [see, for example, 10; 42; 44; 46; 78; 80; 93 -95; 103]. One of the main difficulties in numerical modeling of atmospheric circulation in climate studies and numerical weather prediction is the representation of turbulence in a stably stratified boundary layer. Given the undoubted importance of modeling, we note that only two of the above papers compare model calculations with typical experimental data (i.e., Not with the results of a particular observation), and some articles [e.g., 80] even state directly that the results are direct It is much more efficient to verify the numerical simulation of the VCG using the Large Eddy Simulation models in the atmosphere than using experimental data. As a result, some of the concepts that arise from numerical simulation results are not adequately supported by observations.

To date, there are many experimental data showing the ubiquitous presence of HBV, including VCG, in stable APS. Modern ground-based remote sensing methods - sodars, radars, and lidars - allow visualization of wave-like structures in the APS and determination of their space-time parameters and elevation.

The vast majority of papers published in recent decades on the problem of VKG in the atmosphere are devoted either to the mathematical aspects of the theory of internal gravity waves or to the numerical analysis of hypothetical models. Among the experimental studies on VKG, radar observations of waves in the upper troposphere, lower stratosphere, and ionosphere during regular investigation by radar monitoring networks occupy the main place. There are few VKG observations in the atmospheric boundary layer, apparently explained by the lack of ABL network monitoring and the episodic nature of remote sensing of this layer.

At the same time, IGW studies are very important especially in the atmospheric boundary layer because waves can have a significant impact on exchange processes. Studies of IGW in the atmosphere are necessary to understand and parameterize the atmospheric circulation in numerical regional models and to solve many applied problems.

CONCLUSION

The main results of the work and the following conclusions from it:

1) Methods of sodar registration and determination of Kelvin-Helmoltz wave (VCG) parameters , corresponding to different modes of equipment operation, have been studied. Criteria for visual identification of VCGs on sodar echograms are proposed, showing the dependence of the number of detected wave episodes on the resolution of the equipment.

2) Long-term measurements (over 5 years) of atmospheric boundary layer parameters in the sodar monitoring network in the Moscow region and short-term measurements during expeditions to steppe, arid and coastal regions were carried out. It was found that Kelvin-Helmholtz type gravity shear waves (VKG) are the most common type of waves in weakly heterogeneous terrain and are frequently observed in low level shear flows during reversals.

3) In order to increase the height and temporal resolution of the noise, the instrument and the algorithms controlling the operating mode of the sodar and primary data processing were partially modified. As a result, the number of recorded VCG episodes in the ABL increased by an order of magnitude due to the identification of smaller waves (with periods of about 1 min).

5) The statistics of the observed space-time parameters of the VCG and the characteristics of the associated meteorological profiles were studied using year-round observations at the ZSS. For the period from 2009 to 2011, more than 300 episodes of large VCGs based on soda echograms were recorded at the Zvenigorod scientific station. The duration of VCG packets ranged from 15 minutes to several hours. Basically, waves were observed in the form of oblique bands or "strings" at a height of200 m, with wind shear

3-5 s $^{-1}$ an100 m... Typical parameters of the observed waves: Thickness of the layer of wave activity -60-400 m, spatial period - 0.5-2.5 milesthe average ratio between the wavelength and the layer thickness is L / h = 9.2.The distributions of the observation frequency of the VCR, the period and the length of the observed waves are obtained. The obtained relationship between the wavelength and the layer thickness is generally consistent with the model estimates.

6) The relationship between the observational frequency of the VCG and the synoptic situation has been studied. It is shown that the most frequent long-term episodes of wave activity covering an extended area are formed near pressure troughs and atmospheric fronts at high gradients of the surface baric field (i.e. When the geostrophic wind is strong).

7) The method of obtaining the fine structure of the wind field in periodic turbulent structures from soda data based on a composite analysis, which consists in averaging the intensity of the vertical and horizontal components of the wind speed, measured within individual periods of the VCG. The proposed method of processing soda echo signals using a composite analysis made it possible to obtain the spatiotemporal structure of the wind velocity field in the VCG and to show the relationship between the internal structure and the parameters of the jet flow.

8) Investigated the internal dynamic structure of VCG and its relationship with the average profiles of wind speed. Based on the results of the composite analysis, vector fields of wind speed perturbations within the VCG layer are constructed. Vortex structures typical of Kelvin-Helmholtz instability are found in two-dimensional fields of the velocity perturbation vector. Shown is the opposite vorticity direction of the wind velocity field in a layer with a monotonically decreasing profile of horizontal wind velocity. The revealed structure of the wind field indicates a stable dynamic structure of the VCG represented by slowly rotating vortices (sometimes not closed)

with typical periods on the order of tens of minutes carried by the general flow. The scale and shape of the vortices depend on the shape of the vertical profiles of the wind speed.

LIST OF USED LITERATURE

1. Vasiliev O., Voropaeva O., Kurbatsky A. Turbulent mixing in stably stratified environmental flows: current state of the problem (review) // Izv. RAS: Physics of the atmosphere and ocean. - 2011. t. 47, no. 3. pp. 291-307.

2 Gossard E., Hook W. Waves in the atmosphere. - M.: Mir, 1978.

3. zaitseva D. [et al.] Influence of internal gravitational waves on variations of meteorological parameters of the atmospheric boundary layer // Izv. RAS: Physics of the atmosphere and ocean. - 2018. t. 54, no. 2. pp. 195-205.

4. Kallistratova MA Experimental study of the scattering of sound waves in the atmosphere // DAN SSSR. - 1959 - T. 125 - PP. 69-72.

5 Kallistratova M., Petenko I., Shurygin E. Sodarnye studies of the wind velocity field in the lower troposphere // Izv. Academy of Sciences of the USSR: Physics of the atmosphere and the ocean. - 1987. t. 23, no. 5. p. 451.

6. kallistratova M. [et al.] Sodarny sounding of the atmospheric boundary layer (a review of the work of the Obukhov Institute of Physics and Technology of the Russian Academy of Sciences) // Izv. RAS: Physics of the atmosphere and ocean. - 2018. t. 54, no. 3. pp. 283-300. - DOI: 10.7868 / S0003351518030054.

7. Kamardin A., Odintsov S., Skorokhodov A. Identification of internal gravitational waves in the atmospheric boundary layer using soda data // Optics of the Atmosphere and Ocean. - 2014. t. 27, no. 9. pp. 812-818.

8. Krasnenko NP Acoustic study of the atmospheric boundary layer. - Tomsk: Institute of Optical Surveillance SB RAS, 2001. p. 278.

9. Kuznetsov R. LATAN-3 acoustic locator for the study of the atmospheric boundary layer // Optics of the Atmosphere and Ocean. - 2007. t. 20, no. 8. p. 749-753.

10. Kurbatskaya L., Kurbatsky A. Vortex coefficients of momentum and heat transfer in the upper troposphere and lower stratosphere: a numerical study // Interexpo Geo-Siberia. - 2014. t. 4, no. 1.

11. Lighthill D. Waves in liquids. - M.: Mir, 1981.

12. Lyulyukin V. [et al]. Internal gravitational shear waves in the atmospheric boundary layer according to acoustic site data // Izvestiya Rossiiskoi Akademii Nauk. Physics of the atmosphere and ocean. - 2015. t. 51, no. 2. pp. 218-218.

13 Obukhov, AM, On the scattering of sound in a turbulent flow, Docl. - 1941 - T. 30 - PP. 611-614.

14. Odintsov S. Characteristics of the movements of the lower layer of the atmosphere during the passage of internal gravitational waves // Atmospheric optics. and the ocean. - 2002. t. 15, no. 12. p. 1131.

15. Romanova N., Yakushkin I. On the Hamiltonian description of shear and gravitational shear waves in an ideal incompressible fluid. Izv. RAS: Physics of the atmosphere and the ocean. - 2007. t. 43, no. 5. pp. 579-590.

16. scorer R. Aerohydrodynamics of the environment: per. from English - Peace, 1980.

17. Stepanyants Yu., Fabrikant A. Wave propagation in shear flows // Modern Problems of Physics. Moscow: Fizmatlit. - 1996.

18 Tatarsky VI Propagation of waves in a turbulent atmosphere. - M.: Nauka, 1967 - p. 548.

19. Khomenko G., Panina N. Influence of horizontal wind shears on internal waves in jet streams // Proceedings of the Central Administrative District. - 1977.

20. Shakina N. Hydrodynamic instability in the atmosphere. - Guiderometeoizdat, 1990.

21. argentini S. [et al.] Use of a high-resolution sodar to study surface layer turbulence at night // Boundary Layer Meteorology. - 2011. vol. 143 .-- p. 2011.

22. Baklanov A., Mahura A., Sokhi R. Integrated systems of mesometeorological and chemical transport models. - Springer Science & Business Media, 2011.

23. Banakh V., Smalikho I. Lidar observations of atmospheric internal waves in the boundary layer of the atmosphere on the coast ofSee Baikal// Atmospheric measurement techniques. - 2016. vol. 9, no. 10. pp. 5239-5248.

24. Banakh VA, Smalikho IN Lidar studies of wind turbulence in the stable atmospheric boundary layer // Remote Sens. - 2018. vol. 10, no. 8. p. 1219.

25. Beran DW, Hooke WH, Clifford SF Acoustic echo sounding techniques and their application to gravity wave, turbulence, and stability studies // Boundary Layer Meteorology. - 1973- Vol. 4. pp. 133-153.

26 Flowers W. [et al.]. Turbulence statistics of a Kelvin-Helmholtz wave event observed in the nocturnal boundary layer during the Cooperative Atmosphere-Surface Exchange Study // Dynamics of Atmospheres and Oceans field program. - 2001. vol. 34, no. 2-4. - S. 189-204.

27. Caughey SJ, Reading CJ An observation of waves and turbulence in the Earth's boundary layer // Boundary layer meteorology. - 1975. vol. 9. pp. 279-296.

28. Chapman D., Browning KA Radar observations of wind shear splitting in evolving atmospheric Kelvin-Helmholtz surges // Quart. J. Roy. Meteorol. Soc. - 1997. vol. 123. pp. 1433-1439.

29. Coulter RL, Kallistratova MA The role of acoustic sound in a high-technology era // Meteorol. Atmos. Phys. 1999. vol. 71. pp. 3-13.

30. Cronenwett W., Walker G., Inman R. Acoustic study of meteorological phenomena in the planetary boundary layer // Journal of Applied Meteorology. - 1972. vol. 11. pp. 1351-1358.

31 Cushman-Roisin B. Kelvin-Helmholtz instability as a boundary value problem // Env. Flu. Mech. - 2005. vol. 5. pp. 507-525.

32. Cushman Roisin B., Beckers J.-M. Introduction to geophysical fluid dynamics: physical and numerical aspects. Vol. 101 - Academic Press, 2011.

33 Cuxart J. [et al.]. Stable atmospheric boundary layer experiment inSpain
(SABLES 98): a report // Boundary Layer Meteorology. - 2000. vol. 96. pp. 337-370.

34. DeBaas AF, Driedonks AGM Internal gravity waves in a stably stratified boundary layer // Boundary Layer Meteorology. - 1985. vol. 31. pp. 303-323.

35. deSilvaICH. [et al.] Evolution of Kelvin-Helmholtz waves in nature and laboratory // Earth Planet. Sci. Lat. - 1996. vol. 143. pp. 217-231.

36. Drazin PG The stability of a shear layer in an unbound heterogeneous inviscid fluid // J. Fluid Mech. - 1958. vol. 4. pp. 214224.

37. Drazin PG, Howard LN The instability to long waves of an unbounded parallel inviscid flow // J. Fluid Mech. - 1962. vol. 14. 257-283.

38. Ellrod GP, Knox JA Improvements to an operational index for clear air turbulence diagnosis by adding a divergence trend term // Wea. Forecast. - 2010. vol. 25. pp. 789-798.

39. Emmanuel C. Richardson number profiles through wavebands of shear instability observed in the lower planetary boundary layer // Boundary Layer Meteorology. - 1973. vol. 5, no. 1/2. pp. 19-27.

40. Emmanuel CB [et al]. Observations of Helmholtz waves in the lower atmosphere with an acoustic sounder // Journal of Atmospheric Sciences. - 1972. vol. 29, no. 5. pp. 886-892.

41. Eymard L., Weill A. A study of gravitational waves in the planetary boundary layer by acoustic sounding // boundary layer meteorology. - 1979. vol. 17. pp. 231-245.

42. Feliks Y., Tziperman E., Farrell B. Non-normal growth of Kelvin-Helmholtz vortices in an ocean breeze // Quart. J. Roy. Meteorol. Soc. - 2014. vol. 140. pp. 2147-2157.

43 Fernando HJS, Weil JC Whither the stable boundary layer? // Bull. Amer. Meteorol. Soc. - 2010. vol. 91. pp. 1475-1484.

44. Fritts DC, Wang L., Werne JA Interactions between gravitational waves and fine structure. Part I: Influences of fine structure shape and orientation on flow evolution and instability // Journal of Atmospheric Sciences. - 2013. vol. 70. pp. 3710-3734.

45 Fritts DC [et al]. Stratification accompanying turbulence generation due to shear instability and gravity wave breaking // J. Geophys. Res. 2003. vol. 108, no. 8. p. 8452.

46. fua D. [et al.] An analysis of wave-turbulence interaction // Journal of Atmospheric Sciences. - 1982. vol. 39. pp. 2450-2463.

47 Fukao S. [et al.]. Extensive studies of large amplitude Kelvin-Helmholtz waves in the lower atmosphere with VHF radar in the middle and upper atmosphere // Quart. J. Roy. Meteorol. Soc. - 2011. vol. 137. pp. 10191041.

48. Galperin B., Sukoriansky S., Anderson PS On the critiqueRichardson number in stably stratified turbulence // Atmospheric Science Letters. - 2007. vol. 8, no. 3. - S. 65-69.

49. Gilman G., Coxhead H., Willis F. Reflexion von Schallsignalen in der Troposphäre // The Journal of the Acoustical Society of America. - 1946. - Vol. 18, nein. 2. - S. 274-283.

50. Goldstein S. On the stability of superimposed fluid flows of different densities // Proc. R. Soc. Lond. A, 132 (820). - 1931. - S. 524-548.

51. Gossard EE, Richter JH, Jensen DR Effect of wind shear on atmospheric wave instabilities revealed by FM / CW radar observations // Boundary Laycr Meteorology. - 1973. vol. 4. pp. 113-131.

52. hall FF Acoustic remote sensing of temperature and velocity structure in the atmosphere // Proceedings of the NATO Advanced Study Institute, Statistical Methods and Instruments in Radio Meteorology. - April 1971. pp. 167-180.

53 Helmholtz HV Über diskontinuierliche Bewegungen von Flüssigkeiten // Philos. Mag. - 1868 - Vol. 36 - pp. 337-346.

54. Holtslag AAM [et al.]. Stable atmospheric boundary layers and diurnal cycles: challenges for weather and climate models // Bull. Amer. Meteorol. Soc. 2013. vol. 94, no. 11. pp. 1691-1706.

55. Hooke W., Hall F., Gossard E. Observed the generation of an atmospheric gravity wave by shear instability in the mean flow of the planetary boundary layer // Boundary Layer Meteorology. - 1973. vol. 5, no. 1/2. p. 2941.

56. Howard LN Note on an article by John W. Miles // Journal of Fluid Mechanics. - 1961. vol. 10, no. 4. - S. 509-512.

57. Hunt J., Kaimal J., Gaynor J. Some observations of turbulence structure in stable layers // Quart. J. Roy. Meteorol. Soc. - 1985. vol. 111, no. 469. - S. 793-815.

58. Kallistratova MA Acoustic and radioacoustic remote sensing studies in the CIS (former USSR) // Int. J. Remote Sensing. - 1994. vol. no. 15. pp. 251-266.

59. Kallistratova MA Investigation of low-level jets over rural and urban areas with two sodars // IOP Conference Series: Earth and Environmental Sciences. - 2008. vol. 1, no. 1. p. 012040.

60. Kallistratova MA, Petenko IV Aspect sensitivity of sound backscattering in the atmospheric boundary layer // Appl. Phys. B. - 1993. vol. 57. pp. 41-48.

61. Kallistratova MA, Kouznetsov RD A note on sodar return signals in the stable atmospheric boundary layer // Meteorological Journal. - 2009. vol. 18, no. 3. - S. 297-307.

62 Kaplan ML [et al]. Characterization of severe turbulence associated with commercial aviation accidents. Part 1: A 44-case study synoptic observational analysis // Meteorology and Physics of the Atmosphere. - 2005. vol. 88. pp. 129-152.

63. Kelley ND [et al]. Influence of coherent turbulence on wind turbine aeroelastic response and its simulation: preprint / National Renewable Energy Laboratory (NREL), Golden, CO. - 2005. - NREL / CP-500-38074.

64. kelvin L. Hydrokinetic solutions and observations // Philos. Mag. - 1871. vol. 42. p. 362-377.

65. Knox JA Dynamic meteorology | inertial instability. Encyclopedia of atmospheric sciences. - 2015. pp. 334-342.

66. Knox JA Possible mechanisms of air turbulence in strongly anticyclonic flows // Mon. Wea. Rev. - 1997. vol. 125. pp. 12511259.

67. Kouznetsov RD Multi-frequency soda with high temporal resolution // Meteorological Journal. - 2009. vol. 18, no. 2. - S. 169-173.

68. Kouznetsov RD, Tisler P., Vihma T. Multipoint sodar observations of structures in the ABL over a gently sloping glacier in Antarctica inSummer

2014-2015 // Presentation at the 15th EMS Annual Meeting. -Sofia, Bulgaria, 09/2015.

69 Kouznetsov RD [et al]. Evidence for very shallow summertime katabatic currents in Dronning Maud Land,Antarctica// J. Appl. Meteor. Climatology. - 2013. vol. 52. pp. 164-168.

70. Lapworth A. Observations on the location dependence of the morning wind and the role of gravitational waves in transitions // Quart. J. Roy. Meteorol. Soc. - 2015. vol. 141, no. 686. pp. 27-36.

71. Lawrence GA, Browand FK, Redekopp LG The stability of a shear density interface // Phys. Fluids A. - 1991. - Vol. 3. - pp. 2360-2370.

72nd Little CG Acoustic methods for remote investigation of the lower atmosphere // Proceedings of the IEEE. - 1969. vol. 57, no. 4. - S. 571-578.

73. Liu X. [et al.] Kelvin-Helmholtz surges and their effects on the mean state during gravitational wave propagation // Ann. Geophys. - 2009. vol. 27. pp. 2789-2798.

74 Mashayek A, Peltier WR Shear-induced mixing in geophysical flows: Does the path to turbulence matter for its efficiency // Journal of Fluid Mechanics. - 2013. vol. 725. pp. 216-261.

75. McAllister LG [et al.]. Acoustic sound - a new approach to the study of atmospheric structure // Proc. IEEE. Vol. 57, - 1969. pp. 579-587.

76. Miles JW On the generation of surface waves by shear flows. Part 3. kelvin-Helmholtz instability // Journal of Fluid Mechanics. - 1959. vol. 6, no. 4. - S. 583-598.

77. Miles JW, Howard LN Note on a heterogeneous shear flow // J. Fluid Mech. - 1964, - Vol. 20, - pp. 331-336.

78. Munoz-Esparza D. [et al.]. Bridging the transition from mesoscale to microscale turbulence in numerical weather prediction models // Boundary Layer Meteorology. - 2014. vol. 153 .-- pp. 409-440.

79. Muschinski A. Local and global statistics of clear air Doppler radar signals // Radio Science. - 2004. vol. 39, RS1008. - S. 23.

80. Na JS, Jin EK, Lee JS Investigation of Kelvin-Helmholtz instability in the stable boundary layer using large eddy simulation // J. Geophys. Res. - 2014. vol. 119. pp. 7876-7888.

81.Nappo CJ An introduction to atmospheric gravitational waves. - Academic Press, 2013.

82. Newsom RK, Banta RM Shear flow instability in the stable nocturnal boundary layer as observed by Doppler lidar during CASES-99 // Journal of Atmospheric Sciences. - 2003 - Vol. 30. pp. 16-33.

83rd Odintsov SL Analysis of the microstructure of short-period internal gravitational waves // Continued. 11th Int. Sympos. Acoustic Remote Sensing. - Rome, Italy, 06/2002. pp. 271-274.

84. Optis M., Monahan A., FC Bosveld Limitations and breakdown of the Monin-Obukhov similarity theory for wind profile extrapolation under stable stratification // Wind Energy. - 2016 Vol. no. 19 pp. 1053-1054.

85. Ottersten H., Hardy KR, Little CG Radar and sodar investigation of wave and turbulence in statically stable air layers // Boundary layer meteorology. - 1973. vol. 4. pp. 47-89.

86 Patterson MD [et al]. Time-dependent mixing in layered Kelvin-Helmholtz   surges : experimental observations // Geophysical Research Letters. - 2006. vol. 33, no. 15.

87.Peltier WR, Caulfield CP Mixing efficiency in stratified shear flows // Annu. Rev. Fluid Mech. - 2003, - Vol. 35, - pp. 135-167.

88 Penner I. [et al.]. Detection of aerosol plumes by associated flaring of gas by laser sensing // 21st International Symposium Atmospheric and Ocean Optics: Atmospheric Physics. Vol. 9680. - International Society for Optics, Photonics. 2015. P. 96804D.

89 Petenko I. [et al.]. Wave-like structures in the turbulent layer during morning convection development in dome C, Antarctica // Boundary Layer Meteorology. - 2016. vol. 161, no. 2. - S. 289-307.

90 Plougonven R, Zhang F. Internal gravity waves from atmospheric jets and fronts // Rev. Geophys. - 2014. vol. 52.

91 Romdn-Cascon C. [et al.]. Interactions between drainage flows, gravity waves, and turbulence: a BLLAST case study // Atmospheric Chemistry and Physics. - 2015. vol. 15, no. 15. pp. 9031-9047.

92 Sandu I. [et al.]. Why is it so difficult to represent stably stratified conditions in numerical weather prediction models? // J. Adv. Model. Earth System. - 2013. vol. 5. pp. 117-133.

93. Savijdrvi H. High-resolution simulations of the nighttime stable boundary layer over snow // Quart. J. Roy. Meteorol. Soc. - 2013. vol. 140. pp. 1121-1128.

94. Shapiro A., Fedorovich E. A boundary layer scaling for turbulent catabatic flow // Boundary Layer Meteorology. - 2014. vol. 153. pp. 1-17.

95. Staquet C., Sommeria J. Internal gravity waves: From instabilities to turbulence // Annual review of fluid mechanics. - 2002. vol. 34. pp. 559-593.

96. Stull RB An introduction to boundary layer meteorology. - Kluwer Acad. Publ. p. 666, 1988.

97 Sun J. [et al.]. Overview of wave-turbulence interactions in the stable atmospheric boundary layer // Rev. Geophys. - 2015. vol. 53. - DOI: 10.1002 / 2015RG000487.

98 Sun J. [et al]. Turbulence regime and turbulence intermittency in the stable boundary layer during CASES-99 // Journal of Atmospheric Sciences. - 2012. vol. 69. pp. 338-351.

99. Taylor GI Influence of density fluctuations on the stability of superimposed fluid flows // Proceedings of the Royal Society of London. Series A. - 1931. - Vol. 132, no. 820. - pp. 499-523.

100. thorpe SA experiments on instability and turbulence in a stratified shear flow // J. Fluid Mech. - 1973, - Vol. 61, - pp. 731-751.

101. thorpe SA Transition phenomena and the evolution of turbulence in stratified fluid: a review // J. Geophys. Res. 1987-Vol. 92. pp. 5231-5248.

102. Tyndall J. Sound 3rd // Appleton, New York. - 1875.

103. Udina M. [et al.]. Model simulation of gravitational waves triggered by a density current // Quart. J. Roy. Meteorol. Soc. - 2013. vol. 139. pp. 701714.

104. Venkatesh TN, Mathew J., Nanjundiah RS Secondary instability as a possible mechanism for turbulence in clear air: a case study // Meteorology and Physics of the Atmosphere. - 2014. vol. 126. pp. 139-160.

105. Werne JA, Fritts DC Turbulence and mixing in a stratified shear layer: 3D CH simulations at Re = 24,000 // Phys. Chem. Earth. - 2001. vol. 26. pp. 263-268.

106. Williams A., Hacker J. The composite shape and structure of coherent vortices in the convective boundary layer // Boundary Layer Meteorology. - 1992. vol. 61, no. 3. - S. 213-245.

107 Wyngaard JC Turbulence in the atmosphere. - Cambridge University Press, 2010.

MIX
Papier aus verantwortungsvollen Quellen
Paper from responsible sources
FSC® C105338
FSC
www.fsc.org